Raymond Takes On Climate Change

John Robert Bland

Raymond's great grandfather came across a stranger about to cut down a tree and pleaded with him not to. The beginning of climate change.

Climate change is caused by Global Warming. Global warming is primarily a problem of too much carbon dioxide (CO_2) in the atmosphere—which acts as a blanket, trapping heat and warming the planet. This effect is called the greenhouse effect.

Climate is the average weather. If it snows every December, but last year and this year it is hot like summer; that is climate change. Ok!

As we burn fossil fuels like coal, oil and natural gas for energy or cut down and burn forests to create pastures and plantations, carbon accumulates and overloads our atmosphere.

Earth gets a big headache.

Strange sitings and things happen that just don't make any sense, because the natural order is broken.

And when it got so hot one day that the cement in the street began to crack. Mrs. Johnson got stuck in one of those cracks looking for her key.

Last winter, or was it summer? It got colder than usual and this polar bear showed up at Raymond's front door, or maybe Raymond showed up at the bear's front door.

Raymond works at the big car manufacturing plant, a prime example of pollution that contributes to global warming. He looks awful.

Raymond is very concerned why some people do not believe in global warming when it is chasing them.

Nature is suffering and no one seem to be answering their calls.

It is a constant tug of war, fighting to save the environment against those who are more interested in making money.

The earth was sick and needed urgent care.

It was so bad that leaders from 174 countries came together on Earth Day April 22, 2016, to signed the Paris Agreement on Climate Change. Rejoice! Rejoice!

Raymond was relieved that he may be over those terrible nightmares of running from global warming.

Raymond looked forward to putting it all behind him with a different kind of greenhouse, a cool one.

Raymond was encouraged that the environment had President Barack Obama who got his start as a community organizer.

President Barack Obama was the recipient of the Nobel Peace Prize because he does the right thing.

Something terrible happened in November of 2016.

A new president who was not a friend of the environment and so he pulled the United States from the Paris Agreement on Climate Change. Oh no!

Raymond was sad but like so many others, he got over it to keep on fighting to save the environment to reduce global warming.

Like Denzel, Raymond got stepping and got busy.

Raymond dedicated himself to educate the community to build awareness that will turn into action.

Raymond was delighted that in time people and other stakeholders came together to celebrate a common goal

……and the birds were singing, the trees were smiling, the fishes were playing in water so clear that you could see the bottom, and people were living in harmony throughout the world. It was all because earth got his groove back. He is loved by all.

The End

GLOBAL WARMING

Performed by O'WOW

written by Jone Roparte

Available online at ITunes and CdBaby

I got a thing to say or 2

We talk garbage, non-biodegradable,

While companies wastes are incontainable,

The common-man's actions seem like the only taboo,

Keep everyone's in line, while they're untrainable,

Toxic emissions are keeping them payable,

Fossil fuels not replaced by clean energy use,

I talk too much, it makes no use....

Chorus

Global Warming, don't you think about yourself,

Think environment and someone else,

You can change the world all by yourself,

Change global warming....

Verse 1

Four cars in the driveway

Keeping up with the neighbors

Get rid of the pollution

Do the world a favor

Islands are flooding

Destroying what is in the way

Leaving behind sorrows

What more is there to say

Do the right thing

Don't let this world down

Do the right thing.

Don't let this world down

Verse 2

Meanwhile you buy a SUV

Just because it suits your taste

While on another continent

Children are out there getting your waste

Glaciers in the North Pole

Melting premature

Messing up the food chain

For which you are the cure

Do the right thing

Don't let this world down

Do the right thing

Don't let this world down

Bridge

Do not ask for plastic

Turn down the plastic bag

Get on the bus my friend

You be a hero everlast

Do not ask for plastic

Turn down the plastic bag

Get on the bus my friend

You be a hero everlast

Do the right thing

Don't let this world down

Do the right thing

Don't let this world down

Dedicated

Illustrator Alastair Lard battling Epilepsy